RANKINE CENTENARY LECTURE

RANKINE

His Life and Times

Hugh B. SUTHERLAND, SM, FICE, FIStructE, FRSE

Cormack Professor of Civil Engineering, University of Glasgow

Lecture delivered before the British Geotechnical Society at the University of Glasgow on 13 December 1972 to mark the centenary of the death on 24 December 1872 of *WILLIAM JOHN MACQUORN RANKINE*

LONDON, THE INSTITUTION OF CIVIL ENGINEERS

Published by Thomas Telford Ltd at the Institution of Civil Engineers,
Great George Street, London SW1P 3AA.

ISBN 0 901948 75 6

Made and printed in Great Britain by William Clowes & Sons, Limited,
London, Beccles and Colchester

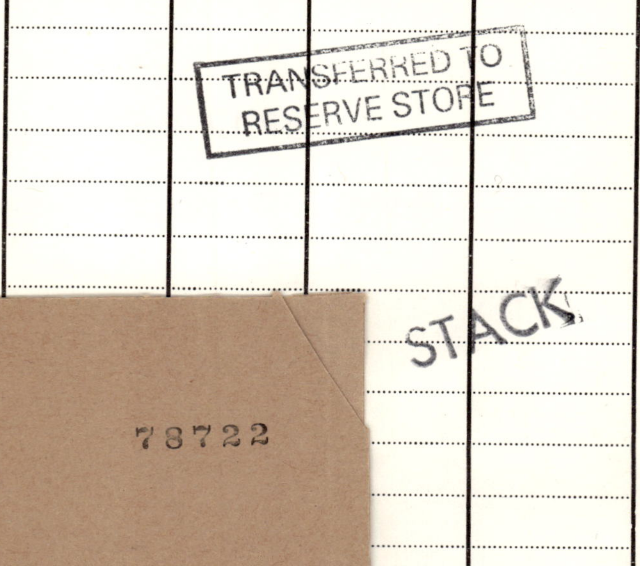

TRANSFERRED TO
RESERVE STORE
STACK
78722

TELEPEN
40 0042812 3

William John Macquorn Rankine, 1820–1872

WHEN I received the invitation some months ago to address the British Geotechnical Society on the subject of William John Macquorn Rankine, I was under no delusions as to the immensity of the task of paying tribute to him, as no length of preparation time would be adequate to review the achievements of this giant among engineers.

Rankine died on Christmas Eve in 1872 at the early age of 52. When he was appointed to the Regius Chair of Civil Engineering and Mechanics at the University of Glasgow in 1855 he was 35, and he had already achieved distinction both as a practising civil engineer and as a mathematical physicist. His seventeen years' tenure of the Chair was a period of intense activity that brought further distinction to Rankine as an individual, and to Glasgow University as an institution of scientific and engineering teaching and learning.

For the past few months I have been apprising myself of his multifarious activities, and I can recommend an exercise of this nature. It is both an exhilarating and a humiliating experience. My reading has led me up many highways and byways, perhaps many more byways than highways, but it has given me a fascinating insight into the world of the academic and also of the practising engineer of 100 years ago.

It has left me with the feeling that anything I can say is a tribute to a giant of the past from a tiny mortal of the present.

Rankine was no ordinary man. He is acknowledged as one of the great pioneers in the effort to bring the resources of mathematics and physical science to the practical problems of the engineer. Since his death a number of tributes have been paid to him by distinguished men of the day, but it is significant that each one of them, even those of Rankine's own time, has taken pains to point out that he did not feel competent to deal with the full range of Rankine's activities, and the lectures have generally been confined to particular aspects of his work.

Little has been recorded of his private life, and in a lecture on the occasion of the centenary of Rankine's birth, Professor Barr[1] deplored the fact that no life of Rankine had been written by any of his contemporaries to give a lasting impression of the man and his times. Practically the only record is a memoir written by his friend Professor P. C. Tait of Edinburgh University and incorporated as an introduction to a volume of Rankine's *Miscellaneous scientific papers* published in 1881.[2] Lewis Gordon, who preceded Rankine in the Chair at Glasgow, and who survived Rankine, was the man who best knew him, and it is unfortunate that Gordon died in 1876 before he had completed his preparation of a review of Rankine's work.

Professor Barr's Centenary Address was followed by an Oration delivered at Glasgow University by Sir James Henderson in 1932.[3] In 1955 Sir Richard

Southwell[4] presented a Memorial Lecture on the occasion of the centenary of Rankine's appointment to the Chair, and Professor James Small has written on a number of occasions on Rankine's work, in particular in 1957[5] on the centenary of Rankine's election as the first President of the Institution of Engineers and Shipbuilders in Scotland. Members of the British Geotechnical Society will no doubt be aware of the essay by Professor Gilbert Cook[6] on Rankine and the theory of earth pressure.

In preparing this lecture I acknowledge my indebtedness to all these previous authors. Most of them wrote in a laudatory vein, apart perhaps from Cook and Southwell, who attempted to apply a critical faculty to some aspects of Rankine's work, particularly in the field of elasticity. Southwell referred to the criticisms of Rankine's work by Todhunter and Pearson,[7] and also to the comments on Rankine by Timoshenko[8] in his volume on the history and strength of materials.

These previous writers on Rankine, with the exception of Cook, were not civil engineers, and this is naturally reflected in the aspects of Rankine's work with which they deal. In a lecture of this nature I must obviously consider the wider aspects of Rankine's contributions, but shall do so fairly briefly. I intend to concentrate on Rankine's role as a pioneer in engineering education, his connexions with the Institution of Civil Engineers, his associations with the practice of civil engineering, and his work in the field of what has come to be known as soil mechanics.

I should also like to try and convey to you what I have gleaned of Rankine's personal characteristics, the atmosphere in which he lived, and the attitudes of engineers towards their contemporaries one hundred or so years ago.

Rankine's life

First of all I should give you a brief summary of his life. Rankine was born in Edinburgh on 3 July, 1820. His father, David Rankine, had been in the army as a young man, holding the rank of Lieutenant in the Rifle Brigade. On leaving the army he took up civil engineering, and after a period with the Edinburgh and Dalkeith Railway he ultimately became secretary of the Caledonian Railway Company. Tait and Gordon, after Rankine's death, had access to Rankine's personal journals wherein it is recorded that Rankine's early tuition in arithmetic, elementary mechanics and physics was obtained from his father. His formal schooling appears to have been brief. He attended Ayr Academy in 1828–29 and the High School of Glasgow in 1830. In that year the family moved to Edinburgh, but the boy Rankine never returned to school because of bad health, and for the next six years his remaining elementary education was carried out by private study at home, supervised and guided by his father. Rankine records that in 1834—he was then fourteen years of age—his uncle gave him a copy of Newton's *Principia* in the original Latin. He read it carefully and this, he states, was the foundation of his knowledge of higher mathematics, dynamics and physics.

At sixteen years of age he entered the University of Edinburgh. This may seem to us a relatively young age to enter university, but in fact this was not unusual when it is remembered that William Thomson (later Lord Kelvin) entered the University of Glasgow at 10, and his brother James Thomson, who succeeded Rankine at Glasgow, was 13 when he started his university career.

Rankine stayed two years at university, and while his stay was short it was brilliant. In the first year he studied chemistry, natural history and botany, and also natural philosophy under Professor Forbes, who was one of the greatest workers of the time in the science of heat. His contact with Forbes at this period probably set Rankine on the road to his becoming, with Clausius and Kelvin, one of the founders of theoretical thermodynamics. In his first year at university he was awarded the Gold Medal for his essay on 'The undulatory theory of light,' and in his second year he gained an extra prize for an essay on 'Methods of physical investigation'. In these two years he also read metaphysics and, as he afterwards told Professor Tait, wasted much time in the study of the theory of numbers. His versatility is shown during these years in Edinburgh by his study of the theory and practice of music—an interest which lasted with him throughout his life, and on which I shall comment later.

It appears that family circumstances then compelled him to leave university and to enter a profession. He became an assistant to his father on the Edinburgh and Dalkeith Railway for a year. He then became a pupil of Sir John Macneill, a leading civil engineer of the day. He had as fellow pupils some others who rose to eminence in the profession—notably Sir Joseph Bazalgette, who was President of the Institution of Civil Engineers in 1883. During his four year pupilage, he was engaged on surveys and schemes for river improvement, water works and harbour works in Ireland, and he also had a spell in the construction of the Dublin and Drogheda Railway.

He returned to Edinburgh in 1842 and worked with railway companies and consultants until about 1848. Sir James Henderson states that around this time Rankine suddenly changed from a practical civil engineer to a theoretical physicist. However, there does not appear to have been such a dramatic transformation. Undoubtedly a much greater part of his time was now to be spent in researches on theoretical physics, but he had been doing work on this earlier, particularly in the field of thermodynamics. However, his output of original work after 1848 was prodigious, and in five years gained for him a Fellowship of The Royal Society. At the same time he was involved in the practice of civil engineering on projects such as the scheme to supply Glasgow with water from Loch Katrine. He also spent a period in London working from the headquarters of Professor Lewis Gordon, who had been the Regius Professor of Civil Engineering and Mechanics at Glasgow since the establishment of the Chair in 1840. The university term only lasted from November to May, and Gordon was required to be in Glasgow only during these months.

Gordon's salary was the princely sum of £275, but he had an extensive professional practice which he conducted from offices in Glasgow and London. Gordon had been dickering with the idea of resigning from the Chair in order to devote more time to his practice. Kelvin indicated that he had heard this in 1847 but Gordon continued in the Chair, although apparently without any great enthusiasm. In fact it appears that the classes in engineering were suspended from 1851 to 1854 and it was towards the end of this period that Rankine was with Gordon in London. Gordon persuaded Rankine to stand in for him in the 1854–1855 session, Gordon having in mind that Rankine should succeed him in the Chair. A copy of the lecture notes that Rankine used during his period as deputy still exists. Gordon's biographer, Constable, says that Gordon delayed his resignation until he was sure that Rankine would succeed him, but the question of succession was not as easily achieved as Gordon perhaps had anticipated.

At this time, John Pagan, the Professor of Midwifery, recommended to the Senate that the Chair of Civil Engineering and Mechanics should be suppressed. The Engineering Chair was a Regius Chair, that is, it had been established by royal decree, but it had not been particularly welcomed by some members of the university as it dealt with subjects far removed from the accepted scholarly pursuits. The Government at this time was endowing the Civil Engineering Chair to the extent of £275 per annum, while Professor Pagan's Chair of Midwifery got but a £50 endowment. There was no doubt good reason for this differential as the sums were fixed after consideration of the fee income from the classes which went to the professor concerned, and the fee income from a small class such as civil engineering would be meagre.

As a slight diversion, it may be of interest to know what the incomes of the Glasgow professors were at this time, and how they were affected by the fee income. Murray[9] in his *Memories of the old College of Glasgow* reports that the Professor of Greek averaged £1663 a year over a five year period, his maximum being £1843. This income, plus a house and no income tax, really put him amongst the wealthy of this time. The other college professors' incomes ranged down to £614 for mathematics, all plus houses.

Pagan's resolution to the Senate was worded, 'notwithstanding the high talents and acknowledged ability of the late Professor, the number of students attending the Class of Civil Engineering had been very small and that the endowment might with propriety be devoted to purposes of greater and more immediate importance in connection with the higher branches of University Education'.

The Senate was fortunately both slow in action and divided in its ranks, and when the Queen's Commission was read on 3 December, 1855, appointing Rankine to the Chair, there was no dissentient voice. At that time an incoming professor was required to write and deliver, as a trial of his abilities, a Latin dissertation. This Rankine did within a week, signed the necessary undertakings and was admitted.

What a loss it would have been to engineering education if Professor Pagan's submission to the University had been accepted.

I should emphasize that good relations between the Departments of Civil Engineering and Midwifery at Glasgow have been restored, as witnessed by the fact that the original Soil Mechanics Laboratory was entered by the Midwifery stair in the old quadrangle. The Department of Midwifery has also been helpful in other ways.

Rankine as engineering educator

During his seventeen years as Professor, Rankine published 111 papers, and I shall comment on this later. Before doing so, however, I should like to discuss his activities as an engineering educator as distinct from his contributions as a physicist and practising engineer. In his inaugural lecture, the translated title of which is 'The Harmony between Theory and Practice in Engineering', and which is published in his *Manual of applied mechanics*,[10] he clearly stated his attitude towards the place of engineering as a University discipline. He noted the three kinds of subjects:

> 'purely scientific knowledge,—purely practical knowledge,—and that intermediate knowledge which arises from understanding the harmony of theory and practice. The third kind is what the engineer is concerned with. It qualifies the student to adapt his designs to which no existing example affords a parallel. Again in theoretical science the question is—what are we to think? But in practical science the question is—what are we to do? We cannot allow our machines and our works of improvement to wait for the advancement of science; and if existing data are insufficient to give an exact solution—that approximate solution must be acted upon which the best data attainable show to be the most probable'.

How pertinent this message is to the present day—particularly in the field of foundation engineering!

Apart from his scientific writings he set about his role of engineering educator in two ways. He soon saw the need for a series of text books on engineering subjects which were based on scientific principles and which were not primarily concerned with contemporary practice, much of which was based on rules of thumb. His output of text books was prodigious. His first and basic text was that on applied mechanics[10] published in 1858. In 1859 a *Manual of the steam engine and other prime movers*[11] appeared. His volume on civil engineering,[12] published in 1862, applied the principles of applied mechanics to the problems of the civil engineer, and a further treatise in 1869 on machinery and millwork[13] was the corresponding manual for mechanical engineers. In 1866 was published his *Treatise on shipbuilding*, *theoretical and practical.*[14] The recognition of the merits of these volumes is that they ran to twenty editions and were translated into French, German and Italian.

Timoshenko[8] in 1953 paid tribute to Rankine's texts and said that even then (almost 100 years after publication) they were not completely superseded. Rankine, through his text books, provided a better basis on which systematic university engineering education could be taught. However, he also had to convince the University and other authorities completely of the suitability of engineering as a university level subject.

In Rankine's time the subject of engineering was attached to the Faculty of Arts, but it was not recognized as a subject qualifying for graduation in Arts. The majority of the early students studying engineering seem to have been men who added this study to the usual course for a degree in Arts. In 1859 Rankine persuaded the Senate to consider the award of a Diploma in Engineering Science. However, the University Commissioners for Scotland stated that in their opinion the University could not confer a distinction other than a degree, and added in any case that engineering was 'not a proper Department in which a degree should be conferred'. Rankine was not discouraged, and obtained approval in 1862 for the award of a Certificate of Proficiency in Engineering Science, an award which existed until 1960.

Even this did not satisfy Rankine, and he pushed for a degree in engineering aided by petitions from his students. In the light of the current demand by students for more involvement in decision making in the universities, it is interesting to note that student participation existed in Rankine's time. Under his pressure, a BSc in science was established in 1872. Thus to Rankine is due the introduction, so far at least as Glasgow is concerned, of the degree in Science in any department of study. It was to take another 50 years or so before a Faculty of Engineering was established, and the Jubilee of that occasion is celebrated in 1973. There is no doubt that any success that was achieved in the 19th century to give engineering an acceptable place among studies in the universities was due to Rankine. Arising essentially out of the recognition of his own abilities and prestige, he set the pattern for engineering education which so many other institutions at home and abroad were to follow.

Rankine as an author

I have referred to Rankine's work as an engineering educator during his term as Professor at Glasgow. These activities in teaching, publishing text books and framing a teaching pattern were sufficient in themselves to ensure him a respected position in any society. These achievements, however, almost pall into insignificance when you consider the number of papers he contributed during the term of his Professorship. The Royal Society *Catalogue of scientific papers*[15] lists 154 publications under Rankine's name. Of these, 111 were published during the period 1855–1872. They ranged over a wide variety of topics and show the all-embracing interests of the man. However, without detracting from Rankine, it should be pointed out that quite a number of these publications were relatively short notes and not papers as we understand them today.

I have always tended to regard Rankine as a civil engineer with a certain amount of civil engineering pride. However, my reading has clearly shown that he was only a civil engineer in terms of the definition of his time, and not in terms of the present understanding of the name.

His published work in what would now be regarded as the field of civil engineering was relatively small. The major part of his work was in thermodynamics, elasticity and hydrodynamics. In the volume of Rankine's *Miscellaneous scientific papers*, edited by W. J. Millar[2], 37 of his papers are included. Only one comes into the category of civil engineering and perhaps surprisingly, his paper 'On the stability of loose earth',[16] which outlined his theory of earth pressure, was not one of those thought worthy of inclusion.

Millar arranged his selection of Rankine's papers in three groups. The first comprised papers on temperature, elasticity and expansion of vapours. The second dealt with energy and its transformations, thermodynamics and the mechanical action of heat in the steam engine, and the third with hydrodynamics, wave forms, propulsion of vessels and stability of structures.

The only paper from the 37 which falls into the category of civil engineering was Rankine's report on the design and construction of masonry dams, published in *The Engineer* in 1872.[17] Rankine had been consulted about a dam to be constructed on the Periyar River in India. He made a theoretical study of the general problem and developed a profile for a dam which almost completely eliminated tensile stress on horizontal sections, the first time that this had been achieved by the application of scientific principles to dam design. In the centenary issue of *The Engineer* in 1956, the late Professor Sutton Pippard[18] praised this work of Rankine. He also described him as one of the greatest of engineering scientists, stating that his influence upon all branches of applied science was to prove incalculable. In the same centenary issue, Professor Sir Arthur Pugsley[19] mentioned the outstanding nature of Rankine's text books, and also drew attention to Rankine's contributions on arch design.

Timoshenko[8] paid tribute to Rankine and his presentation of the mathematical theory of elasticity. He states that Rankine was the first person to define stress and strain rigorously. Lord Kelvin also acknowledged Rankine's definition of these terms in a paper on the theory of elasticity published in 1855.[20] Todhunter and Pearson[7] commented favourably on Rankine's definitions but sought to improve them by adding the term *sprain* for the strain produced by a wrench. 'Sprain' seems a peculiar term, but I presume it sounds no more peculiar now than did the term 'strain' when it was first introduced. Timoshenko also noted that Rankine's method of parallel projections in statics suggested the famous work on reciprocal figures to Clerk Maxwell.

Clerk Maxwell,[21] in turn, in an article in *Nature* in 1878, classed Rankine along with Kelvin and Clausius as 'the three founders of thermodynamics'. Of these three pioneers, Maxwell credited Rankine with availing himself to the greatest extent of the 'scientific use of the imagination,' and points out the advantages that Rankine had in this connexion through his accomplishments as an engineer. He also commented that the scientific career of Rankine was 'marked by the gradual development of a singular power of bringing the most difficult investigations within the range of elementary methods'. However, regarding his earlier papers, Maxwell states that 'Rankine appears as if battling with chaos, as he swims, or sinks, or wades, or creeps, or flies.' He concluded by saying that Rankine's premature death was as great a loss to the diffusion of science as it was to its advancement.

In view of these laudatory comments on Rankine by distinguished engineers and scientists it is interesting to consider the comments of Sir Richard Southwell[4] in his 1955 lecture. Southwell emphasized the obvious difficulty of assessing the worth of any man's work in the light of knowledge one hundred years on. From his own judgment he was only prepared to comment on Rankine's contributions to elasticity, and he concluded that he could find no outstanding contribution which still has influence on our thinking. He criticized the obscurity of Rankine's papers, and quoted Todhunter and Pearson[7] to the effect that Rankine lacked the mathematical equipment demanded by the complexity of his notions. What Pearson actually complained about was that Rankine, in a paper in 1855,[22] had obtained his results using 'that branch of the Calculus of Forms incorporating the recent researches of Sylvester, Cayley, and Boole.' Pearson complained that 'it would rarely happen that the elastician will have made a sufficiently wide study of invariants, covariants, contragredients, etc., to understand the processes of this Memoir while terms such as Umbral matrices and contra-ordinates tend at least to obscure the simple physical principles which often lie behind the equations.' He commented that St Venant presented shorter and more direct proofs of Rankine's equations eight years later in 1863. Professor Small[23] also drew attention to Lord Kelvin's oft-quoted statement that 'I never satisfy myself until I can make a mechanical model of a thing' mentioned in one of his Baltimore lectures, and in which he turned aside to express some criticism of Rankine for not having adopted the same attitude to a problem. Professor Small suggests that the mathematical physicist, Kelvin, (as did the mathematician, Pearson), found the engineer Rankine too abstruse!

Southwell's comments on Rankine were mainly based on those of Todhunter and Pearson, but it should also be noted that these two authors were also critical about other great figures in elasticity and the strength of materials. What man can have all his writings gone through in minute detail without a misconception or error being exposed? I have been informed that Kelvin thought that there was no future in alternating current. I well remember it being put to Terzaghi that he had changed his view on an issue. His devastating retort was that if he could not change his mind he might as well be a fossil.

It has always been the same. The giants such as Rankine have had the ability, energy and courage to present their views on a subject, or in Rankine's case, many subjects. Without this calibre of man, and they exist in our time, the critics would not have anything to criticize, and where would we be?

In the late 1850s Rankine apparently started taking an increasing interest in problems of hydrodynamics and ship performance. From this time to his death he published about 30 papers and communications on these topics. This interest probably developed through his contacts with engineers and shipbuilders on the Clyde and through the creation of the Institution of Engineers and Shipbuilders in Scotland in 1857, an Institution of which he was the first President.

It is to be remembered that in the midst of his theoretical work Rankine was having to deal very actively day to day with Clydeside engineers and shipbuilders, who prided themselves on being 'practical' and who would not attempt to understand his major work. To get things done he had to be continually writing and arranging affairs with them in the simplest possible way, and this must have made severe inroads on his time.

One example of such collaboration is recorded by Tait.[24] In 1857 J. R. Napier provided Rankine with experimental data on the engine power required to propel different types of steamship at various speeds. From these data Rankine deduced general formula for the resistance of ships. The work was done in confidence and applied with success by Napier in his business. Rankine was not permitted to publish, as Napier wished to take full commercial advantage of the work. However, in order to place his new law in recorded form, Rankine[25] submitted a letter to the editor of the *Philosophical Magazine*, which contained a statement of the 219 letters of the law in anagram form. No one apparently solved the anagram until his papers were consulted after his death.

Napier commented upon the high quality of Rankine's mind, and the rapid and keen insight which took him immediately to the heart of a problem.

Rankine published many papers in the *Transactions* of the Institution of Naval Architects, one of which in 1865[26] is stated to be the first paper in English on the theory of propellors. These papers, along with the *Treatise on shipbuilding*, give Rankine a leading place in the field, as was recognized by C. W. Merrifield, Honorary Secretary of the Institution of Naval Architecture, who in 1872 said 'I think few men now living have done more for the theory of ships, taking it altogether than Professor Rankine.'

And I think, this expression of Merrifield's, viz. 'taking it altogether' is the only way in which any abiding influence a man has can be assessed.

Rankine as civil engineer

Rankine had a rather tenuous connexion with the Insitution of Civil Engineers. He was elected to the grade of Associate at a meeting on March 1843, at which His Royal Highness Prince Albert was elected by acclamation as an Honorary Member. At that meeting Rankine[27] submitted a paper on railway axles and the means of preventing accidents by observing the law of continuity in axle construction. He showed for the first time the danger of abrupt changes in the transverse dimensions of machine parts, and also the danger of the interruption of the fibre of wrought iron parts forged with sharp re-entrant angles. These observations by Rankine preceded St Venant's work on re-entrant angles by 12 years.

A week later, at a meeting on 14 March, he presented a paper[28] on a method of setting out railway curves, which subsequently became known as Rankine's method. At this meeting he also introduced a paper[29] on a contrivance which had been devised by his father, Lieutenant D. Rankine, to allow the springs of railway carriages to give an easier ride under different loadings. For these papers Rankine was awarded a Walker Premium.

James Walker at this time was the President of the Institution of Civil Engineers, and in his presidential address[30] in January 1842 he made some pertinent comments about the education and training of civil engineers in his day. Some indication of the possible employment difficulty for junior civil engineers at this time is conveyed by Walker's reference to young civil engineers who, having finished their apprenticeship, were out of employment. He recommended that until such times as they could be engaged by a professional engineer, they could be usefully, though not necessarily profitably employed, with the metalwright, the joiner, the ironfounder, or the general builder. He also noted that the Government of the country paid little attention to the cultivation of the science of engineering, a notable exception being their endowment of a Professorship of Civil Engineering at Walker's own university, the University of Glasgow.

I mention Walker as his name is one that is well known to the Faculty of Engineering at Glasgow, although I did not appreciate the connexion until I was preparing this lecture. He inaugurated the Walker Prizes in Applied Mechanics which are still awarded annually. Rankine announced their inception in a lecture to the students in 1856.[31] Two were to be awarded annually, one by the examiners, and one by the vote of the class, the competitors included. Another example of student participation of one hundred years ago!

In 1844 Rankine[32] read a paper to the Institution on a safety drag that had been devised to prevent trains running away on inclines. This arose through his work with the Edinburgh and Dalkeith Railway Company, and another paper followed in 1848[33] dealing with the effects of severe storms on sea walls. As a young man in his teens, Rankine had been resident engineer for the railway company on the construction of the walls, and the alterations he made to the original designs showed a high degree of thoughtfulness and confidence for one so young.

His contributions to the Institution were brief from now on. He participated in a discussion on the Keyham Dockyard on 9 May, 1854, a meeting at which a lady, Lady Bentham, was allowed to contribute through the courtesy of the Secretary, and then in the annual Report of the Institution for 1857–1858 there is a note to the effect that Rankine had resigned from the Institution as an Associate. A bit of mystery surrounds this, but it would appear that he had an argument with the Institution through their failure to transfer him to the grade of Member, even though he had held the Chair at Glasgow for two years. Despite his severance of the link with the Institution, he subsequently contributed to a few discussions, but his interests seemed to have moved away from civil engineering, although in his last contribution to the Institution *Proceedings* in December 1870 he participated in a discussion on metal and timber arches.

I have tried to find further information on Rankine's work as a civil engineer in practice, but with no great success. We know a little of his work as a railway engineer, and of his collaboration with his father in both railway projects and in associated research, but the only other record of a civil engineering project I have been able to trace is his connexion with a scheme for the improvement of the water supply to Glasgow.

In the early 1800s in Glasgow, increasing dissatisfaction was being expressed about the quality and quantity of the water supply for the city. The main source of supply was water pumped from the River Clyde, and the increasing pollution of this source, allied to the limited quantities of water available, led to extensive investigations and proposals for an alternative source of supply. The provision of water was in the hands of a private company called the Glasgow Water Company, which apparently fought a delaying action for a period of 20 years or so with the citizens of Glasgow, who were determined that some improvement should be made. Lewis Gordon, while Professor of Civil Engineering, entered into the controversy in 1845, and his submissions and those of Rankine at a later date give an interesting insight into the public role that engineers were prepared to play at that time, and also the critical attitude that engineers were prepared to take publicly towards one another—a publicly expressed attitude that would rather be frowned upon nowadays, or in fact even disallowed by the current professional code of ethics.

In 1845, Gordon, in association with Laurence Hill, addressed a report to the chairman and the provisional committee of the Glasgow Gravitational Water Company in which they put forward proposals for providing a new water supply to Glasgow from Loch Katrine, which is about 40 miles from the city. Their scheme provided for aqueducts, the construction of a storage reservoir near Glasgow and the distribution system within Glasgow, at a total cost of £400 000. Apparently Gordon and Hill had developed their scheme on their own initiative and at their own expense. Their report was enthusiastically received, a provisional committee of 34 distinguished gentlemen was formed, and under their authority a prospectus was issued for the formation of a new company, the Glasgow Loch Katrine Water Company, with a capital of £500 000 pounds. The shares of this proposed company were eagerly taken up by the public, and preliminary arrangements were made for bringing the scheme before Parliament. However, the existing Glasgow Waterworks Company felt that their interests were being encroached upon, and they in turn put forward a scheme for providing water from Loch Lubnaig, also to the north of Glasgow.

The publication of the Loch Lubnaig scheme led the provisional committee of the Loch Katrine scheme to believe that this was a *bona fide* movement in the right direction, and consequently further steps towards establishing the new Loch Katrine Water Company were stopped, the deposits in the shares were returned to the subscribers, and the professional men, namely Lewis and Gordon, relinquished all claim for the time and expense they had devoted to establishing the principle that Glasgow could be supplied with pure water from Loch Katrine at a very moderate expense. In 1846 an enabling act was obtained to bring water from Loch Lubnaig. This scheme, however, as laid out by the Glasgow Water Company, proved to be impracticable from cost considerations and also because of the engineering difficulties involved, and no part of it was put into effect. The matter appeared to lie in abeyance for some years until Rankine and John Thomson, a civil engineer in Glasgow, entered the lists.

In a letter *cum* report[34] dated 1 March, 1852, addressed to the Lord Provost, Magistrates and Council of Glasgow, Rankine and Thomson again raised the matter of using Loch Katrine for the water supply of Glasgow. Rankine indicated that he thought the Loch Lubnaig scheme was a subterfuge adopted by the Glasgow Water Company to bring disrepute on any proper gravitational scheme such as the Loch Katrine scheme. He made the statement that the Loch Lubnaig scheme had been laid out not for execution, but merely as an obstacle to the success of better projects. Rankine then went on to outline the benefits and advantages of the resuscitated Loch Katrine scheme, deliberated on the quality and quantity of the water available, and gave a breakdown of financial costs involved and the amount of water rate that would have to be levied. He gave the capital costs of construction as £460 000, and advocated that the undertaking could be carried out on a profitable basis by a commercial company with participation by the Corporation in the form of the taking up of debentures or by a provisional loan. Four weeks later, Rankine and Thomson followed this first report with a second letter[35] in which they further considered the relative merits of the Loch Katrine and Loch Lubnaig schemes, argued for the supremacy of the Loch Katrine scheme, and confirmed the estimate of costs that they had submitted in their earlier report.

I can only surmise what happened in the following months. The Corporation of the City of Glasgow seem to have taken the initiative. They sought a Bill in Parliament to purchase the works and interests of all the water companies who were supplying water to the Glasgow area, and in 1853 they appointed Mr John Frederick Bateman, civil engineer of Manchester and London (President of the Institution of Civil Engineers in 1877), to report to the Glasgow Town Council on the feasibility of the Loch Katrine scheme. Rankine's two communications of 1852 were probably written in the knowledge of the action the Town Council were about to take in 1853. It would appear that he was bringing to the attention of the Town Council the previous work on the Loch Katrine scheme, prepared by Lewis Gordon and Hill.

Rankine's support for Gordon had therefore gone unheeded by the appointment of Mr Bateman, and this in turn led to a detailed and impassioned letter and report by Lewis Gordon to the Lord Provost. This was quite an unusual public submission to be made by an eminent consulting engineer. In fact it was headed as 'our respectful remonstrance', but it was in effect a strong public protest against Bateman's appointment, Gordon's arguments being based not only on the fact that he claimed to be the originator of the scheme, but also that Mr Bateman proposed to involve the Corporation in an outlay of twice the amount which was ascertained and publicly stated by Gordon in 1845 and confirmed by Rankine in 1852. He strongly submitted that Bateman's costs were a 'most unnecessarily exaggerated estimate', and Gordon stated that 'he was willing to stake his professional reputation to be able to demonstrate this fact.' In concluding his letter, he said, 'I venture to add to this remonstrance a word of advice at the risk of its being treated as is proverbial with advice not asked for', the word of advice being to the effect that a municipal authority were generally incapable of efficiency in supplying water to the community, or dealing with any other matters of interest to the community, and that by far the best way of conducting these operations was to put them into the hands of a joint stock company.

I wonder if Gordon really thought that his protest would bring any action to his own benefit? I doubt it, in view of his slighting comments on the efficiency of municipal authorities, but the end result was that his plea was ignored. After some further delays in Parliament, the Loch Katrine scheme was authorized to go ahead, following further reporting by Robert Stevenson and Brunel. Mr Bateman was confirmed as the engineer, and the scheme was opened in 1859, the total cost being £920 000, which in the end, was much greater than Gordon, Rankine and Bateman had originally estimated—a financial outcome that is not confined to the 19th century.

Rankine on soil mechanics

I should now like to consider the background to Rankine's[16] paper on earth pressure theory and to consider the state of knowledge and teaching that existed at that time in this subject. Southwell tried to throw some light on the matter by suggesting, and I quote, 'that Rankine's reading for the most part concentrated on British work in his field of interest, and that this was not an uncommon tendency in his time'. Before I read Southwell's suggestion, I had been reading the early *Proceedings* of the ICE and other publications, and had formed the impression that our predecessors were extremely well informed on procedures and developments in other countries, particularly having in mind how much more difficult communications were then than at present. However, I shall consider Southwell's comment as it particularly relates to Rankine and Lewis Gordon.

Lewis Gordon was a most interesting and capable character. As a young man he spent some years in Germany as a student at the Royal Academy of Mines at Freiberg under Julius Weisbach, the Professor of Mechanics and Applied Mathematics. From Germany, Gordon went on to France, where he spent some time at the Ecole Polytechnique. In 1847[36] Weisbach published two volumes entitled *Principles of the mechanics of machinery and engineering*. In less than a year both these volumes had been translated into English and published in London, Gordon being responsible for the translation of the second volume[36] of 420 pages.

In 1847 Gordon[37] published a 100 page booklet giving a synopsis of the lectures to be delivered to the students of civil engineering and mechanics at the University of Glasgow in the 1847–48 session. His subtitle was *Engineering aphorisms and memoranda*, and the booklet contained not only an account of his lectures, but also the reading list associated with them. His section on earth pressure included references to Coulomb's original paper, to other French papers by Poncelet, Prony, Garides, Mayniel and Collin, to German papers by Hagen and Kozsegh, and to British works by Pasley and Moseley.

In this booklet Gordon gave an expression for the total earth pressure exerted by a cohesionless soil against a vertical wall which leads to pressures of about two-thirds of those obtained by the later Rankine theory. Gordon recognized cohesion but left it out of consideration (as he noted, 'as is usually done') and also stated that he 'neglected the friction of the soil on the back of the wall,' justifying this by stating that 'this friction does not come into play till the wall is in motion under the influence of a pressure normal to the wall and which is quite independent of friction. It is only then that the earth begins to sink and that the influence of friction manifests itself'.

He advocated that the pressure so obtained should be increased by 15%, and criticized as extravagant the French engineers who applied an increase of 50% to the calculated value. This is an interesting comment, as the French engineers' design values of earth pressure would have led to values approximately equal to those obtained by the Rankine theory.

As has been stated, Rankine deputized for Gordon before he succeeded to the Chair, and he would almost certainly be familiar with the foreign references Gordon recommended to his students. Rankine was also familiar with Moseley's book,[38] which was published in 1843 and which gives a detailed account of the wedge theory of earth pressure, working through the cases of total active and passive pressures for walls with and without wall friction. Moseley also recognized the existence of cohesion, but advocated its neglect as he considered it would deteriorate with time. It is interesting to note, as Golder reports in his essay on Coulomb,[39] that Coulomb also advocated the neglect of cohesion in earth pressure calculations, his reason being that newly deposited earth possesses no cohesion (the opposite reason from Moseley and Gordon).

Rankine would also be aware of Gordon's translation of Weisbach who has a detailed chapter on earth pressure. Weisbach considered active and passive pressures based on wedge theory for cohesionless soils with no wall friction. Wisbach also developed the case for cohesive soils, determined the height to which a vertical face can stand unsupported, and obtained expressions for total active and passive pressure for cohesive soils identical to those presented by Bell in 1915.

What I am seeking to demonstrate is that Rankine would be familiar with the earlier work on earth pressure when he took up the Chair in Glasgow, although he may not have given much detailed thought to the subject until he had to prepare the curriculum for the course. This would appear to be the case, for in a letter[40] to the Secretary of the Royal Society dated 18 February, 1856, he stated that in the course of preparation of his lectures he had 'found it necessary to re-investigate the Mathematical Theory of the Stability of Earthwork and Masonry', and indicated that he was preparing a paper on the subject for submission to the Society. The paper duly appeared in the *Philosophical transactions* in 1857 entitled 'On the stability of loose earth'.[16]

Rankine in his paper and subsequent books stated that previous researches on earth pressure were based on some mathematical artifice or assumption, such as Coulomb's wedge of least resistance. He went on to say that researches so based, although leading to true solutions of many problems, were both limited in the application of their results, and unsatisfactory from a scientific point of view. He proposed, therefore to investigate the mathematical theory of the frictional stability of a granular mass, without the aid of any artifice or assumption, and from the following sole principle:

> 'The resistance to displacement by sliding along a given plane in a loose granular mass is equal to the normal pressure exerted between the parts of the mass on either side of that plane, multiplied by a specific constant.'

This specific constant he called the coefficient of friction of the mass, and is the tangent of the angle of repose.

He also discussed cohesion and recognized its effects, but he concluded that it was gradually destroyed by the action of air and water and could not therefore be relied upon to provide permanent stability.

Rankine gave no specific criticisms of Coulomb's approach, although he does refer to it as a mathematical theory of the combined action of friction and adhesion in earth, but he said 'for want of precise experimental data, its practical utility is doubtful'. This appears to be the case of the pot calling the kettle black, as there is no record of Rankine attempting to justify his own theory by experiment! I do not think he paid a great deal of attention to Coulomb's work, but was more intent on extending to granular materials his more general treatment of the states of stress within solids using the ellipse of stress concept.

On the basis of the conjugate stress concept he developed equations based on 'principles which are common to the internal equilibrium of a solid mass in what manner soever constituted'. In applying these equations to an 'incoherent granular mass' Rankine then introduced what he described as the 'peculiar principle' which he had stated at the beginning of his paper, namely that for stability of this granular mass the greatest obliquity of pressure at any point in any plane must not exceed the angle of repose.

Gilbert Cook has examined Rankine's original paper in some detail, and I refer readers to this excellent review.[6] I endorse Professor Cook's comment that few readers of the paper can have found it easy to follow and understand completely. The mathematical formulation is abbreviated in a manner that on occasion it is merely a succession of statements, the connexion between which it is difficult or impossible to detect. The notation he uses is obscure even by the standards of his times, and does not assist in an easy understanding.

Rankine's contribution to earth pressure theory would, in my opinion, have remained virtually unknown if he had not followed it up with his series of text books. The first of these volumes was his *Manual of applied mechanics*. His presentation therein of his theory of earth pressure, while it contained the same material and conclusions as in the original paper, was in a much more readable form. He used a notation for stress which has survived to the present day and applied his theory to the design of retaining walls in a manner that is still given in recently published text books. He incorporated much of the same material in his *Manual of civil engineering*, warning his readers that the properties of earth with respect to adhesion and friction are so variable that the engineer should never trust to tables or to information obtained from books to guide him in designing earthworks, when he has it in his power to obtain the necessary data either by observation of existing earthworks in the same stratum or by experiment. Nevertheless on the preceding page in his book, as shown in Table 1, he gave a table of angles of repose for all types of soil, including peat and clay of different consistencies.

In the period of 20 years or so after Rankine's publications, there appears to have been a hiatus in Britain in papers on earth pressure on soils. There was some activity in France, however, by Boussinesq and his co-workers. In the United Kingdom this was probably a time of application of the theories and of digestion and reflection on the results.

However, the forces of reaction were mounting, and the first real assault on the Rankine and other earth pressure theories was made by Sir Benjamin Baker in his celebrated paper in 1881 to the ICE, entitled 'The actual lateral pressure of earthwork'.[41] This paper gave birth to one of the most often used quotations in soil mechanics, that Baker 'considered that the deductions from earth pressure theory stood really on exactly the same scientific basis, and the same practical value, as the weather forecasts for the year in Old Moore's Almanack.'

Baker was really attacking the Rankine theory, and illustrated his argument of the conservative nature of that theory for cohesionless soils by a number of examples, one being the case of a pile of rubble retained by a vertical wall of pitch pine blocks. The discussion and correspondence on the paper were most interesting. Flamant and Boussinesq rushed to the defence of theory—not Rankine's, I may say, but their own. Flamant applied the Boussinesq theory to the same example as Baker, pointed out that the deficiency in the Rankine theory was the neglect of friction on the back of the wall, and then showed that Boussinesq theory, when applied with the angle of wall resistance equal to the angle of repose, indicated that Baker's pitch pine block wall was stable. Flamant's calculated value of earth pressure was, incidentally, identical with that of Coulomb for these same soil conditions.

There was a further rallying to the cause of the theorists by Professor Gaudard in the correspondence on Baker's paper, a defence which has been reiterated on various occasions since 1881. It is interesting to look at Gaudard's remarks to the effect that 'although the Author (Baker) had treated the theorists a little severely, he had not pierced their armour, neither had he himself furnished sufficient light; he had simply demonstrated—what no one doubted—that no theory could be applied to a case outside the conditions for which the theory was itself correct, the more so when it was reduced to a maimed condition'.

Table 1. Angles of repose given by Rankine

Earth		Angle of repose, ϕ
Dry sand, clay and mixed earth	from	37°
	to	21°
Damp clay		45°
Wet clay	from	17°
	to	14°
Shingle and Gravel	from	48°
	to	35°
Peat	from	45°
	to	14°

It is worthy of note, however, as Cook indicated, that when experiments were conducted on material in the conditions assumed by Rankine, i.e., with wall effects eliminated, Rankine's predictions were confirmed.[42]

While there were some further contributions on earth pressure in the early years of the 20th century, notably by Bell, Terzaghi[43] in 1936 was the first person to examine closely the validity of earth pressure theory. In his paper Terzaghi was primarily concerned with the effects of strain and deformation on the distribution of earth pressure, and not so much with the evaluation of total earth pressure.

He pointed out that the fundamental assumptions of Rankine's earth pressure theory are incompatible with the known relation between stress and strain in soils, including sand. Therefore the use of this theory should be discontinued, particularly in the strutted excavation type of problem. However, he also said that the Rankine theory covered a broader field of application than Coulomb's.

He concluded with the admonition to designers, as Gaudard had done in 1881, that 'the fundamental misconception associated with the traditional earth pressure computations does not reside in the theories as such. It lies in the failure of the designers to consider the limitations on the validity of the theoretical results'. Cook and Southwell reiterated Terzaghi's comments and strictures almost as absolution for Rankine, but I wonder if Rankine can be held entirely blameless in the matter, a point I should like to consider a little further.

I have commented earlier on Rankine's inclusion in his *Manual of civil engineering* of a table of angles of repose for all types of soil including peat and clay. In the *Manual* he at least stated the conditions to which his theory applied and gave cautionary notes on the application of earth pressure theory. However, the publication of his table may have started the accelerating process which apparently led to the almost blind application of his theory to all problems in earth pressure.

Rankine followed up his *Manual of civil engineering* with a volume published in 1866[44] called *Useful rules and tables*, the object of which was 'to provide, in moderate bulk, a collection of Rules and Tables relating to those parts of mathematical and mechanical science whose application most frequently occurs in the useful arts, and especially in engineering and practical mechanics. The use of algebraic symbols is avoided, except in those cases in which the rules cannot be clearly expressed without them.'

Here was what the engineers of the day were probably waiting for—320 pages for 45p—the bane of the existence of creative thought—the panacea for all your problems—look up page so-and-so and substitute in the formula. How often have we all fallen into the convenience of this approach, sometimes oblivious to the consequences!

I have already made reference to Rankine's inaugural lecture delivered in Latin to the Senate of the University. In this lecture he distinguished between purely scientific instruction and instruction in practical science, and drew attention to the different type of mental faculty that the engineer has to possess compared with the scientist.

'In theoretical science, the question is—what are we to think?' And he emphasized the time that may be required to solve the question. 'In practical science, the question is—what are we to do? In doubtful cases, a prompt and sound judgment is one of the characteristics of a practical man, in the right sense of that term.'

Having these remarks of Rankine in mind, it is interesting and self-revealing to see how civil engineers have handled the Rankine and other theories of earth pressure.

The relative scientific merits of these theories have been discussed and commented upon on many occasions by Terzaghi and others. Nevertheless it is interesting to look at the treatment given to earth pressure theory in the most recent text books from both sides of the Atlantic. Both Rankine and Coulomb theories are treated in some detail, their strengths and weaknesses discussed, and things are pretty well left at that, i.e., almost the attitude that Rankine's theoretical scientist would adopt.

Terzaghi's[45] approach, however, is different. Here we see a man, in full awareness of the limitations of earth pressure theory from a scientific point of view, adopting Rankine's approach of the practical man at work, by giving his recommendations as to what we are to do rather than what are we to think.

Terzaghi's method of estimating the earth pressure on what he describes as small retaining walls (although he does not specify what constitutes a small retaining wall), was to classify backfill into five types, type 1 being clean sand and gravel, the material to which the Rankine and Coulomb theories, as originally conceived, can be taken to apply. Terzaghi appears to have chosen a ϕ value of 33° for his type 1 soil, and a comparison can be made between Terzaghi's values for k_h, which is a measure of the horizontal component of earth pressure, with those obtained from the Rankine and Coulomb theories. Fig. 1 shows that there is relatively little difference between the results from Rankine's theory and Terzaghi's recommendations.

Terzaghi later states that his semi-empirical procedure gives greater earth pressures than those obtained from earth pressure theory, but there is not a great deal in it for the case I have illustrated. Terzaghi goes on to emphasise that the design of a retaining wall on the basis of theory is justifiable only if the physical constants of the backfill material are reliably known, and also if provisions are made to ensure that the porewater pressure in the backfill will permanently be negligible. The cost of satisfying these requirements offsets the benefits obtained from the use of theoretical earth pressure in design, unless the retaining wall is of greater than ordinary height or length. In this event, however, it may prove more economical to investigate the properties of the backfill, to take suitable measures to ensure that the properties will remain constant, to eliminate the possibility of excess porewater pressure,

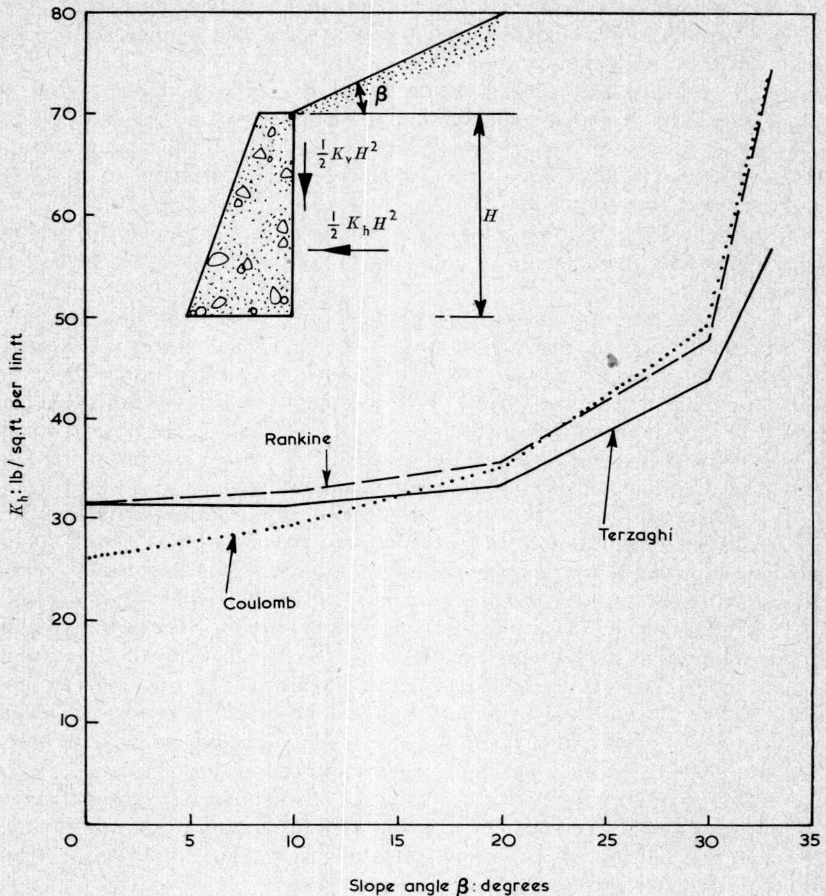

Fig. 1.

and to design the wall to withstand only the theoretical value of the earth pressure.

It is interesting to note that Rankine in his *Manual of civil engineering* emphasized the same sort of points as Terzaghi. He drew attention to the methods to be used when placing the backfill to retaining walls, and how adequate drainage should be provided behind and in front of walls.

An interesting sidelight on compaction of fill material for embankments is Rankine's[12] reference to a Mr Smith's method of placing puddle clay core material. Mr Smith rammed and puddled each successive layer of a reservoir embankment by erecting a rail-fence along each side of it, and driving a flock of sheep several times backwards and forwards along it. The forerunner of the sheepsfoot roller!

There has not been much reported about Rankine's work on the bearing pressures that soils can sustain, although in his original paper[16] a short section was devoted to the application of his earth pressure theory to the stability of foundations, and in particular to the minimum depth to which a foundation should be placed in the soil. He included this work on bearing pressures in his *Manual of civil engineering* and also in his book *Useful rules and tables*, and it is interesting to look at the bearing pressures that were used in Rankine's time. In his *Manual* he gave *examples* of the intensities of pressure that had been applied to rocks and soils. When these values appeared in his *Useful rules and tables* they could, and probably would have been read as *advocated* intensities. Table 2 gives his values for what he classed as first or second class foundation materials. His third class of material was described as soft or loose earth. For a uniform pressure p, applied to a foundation on class 3 material, the minimum depth of the foundation was given as

$$\frac{p}{\gamma}\left[\frac{1-\sin\phi}{1+\sin\phi}\right]^2$$

where γ was the unit weight of the soil and ϕ its angle of repose.

Rankine never commented on the fact that this expression leads to zero bearing capacity for a foundation on the surface of a cohesionless mass. Nevertheless, it was probably the first attempt to develop an expression for bearing capacity, and was the basis of the formula which Bell later developed for cohesive soils. I mentioned earlier Southwell's suggestion that British engineers of Rankine's time were unfamiliar with work outside the UK. My impression is that the European engineers were even less familiar with Rankine's work. I have come across a number of examples of this, one of them being Timoshenko's reference to a Russian paper by Pauker[46] published in St Petersburg in 1889 which proposed a theory for determining the necessary depth of a foundation identical to that derived by Rankine 33 years previously. To verify Pauker's (and Rankine's) formula, Kurdjumoff[47] carried out tests

Table 2. Values of bearing pressure given by Rankine

Loads on ordinary foundations	
First class	*Ton/sq. ft.*
Rock, moderately hard, strong as the strongest red brick	9·0
Rock, of the strength of good concrete	3·0
Rock, very soft (crumbles in hand)	1·8
Second class	
Firm earth; hard clay; clean dry gravel clean sharp sand, prevented from spreading sideways	from 1 to 1·5

on sands and reported that a greater pressure than that predicted by the formula was required to cause failure.

I often wonder what Rankine would have thought about all that has been written and debated on his theory of earth pressure. There seems little doubt that his paper was written to satisfy an intellectual whim. Apart from including it in his text books he never returned to the problem from either a theoretical or practical point of view. Nevertheless, despite the legitimate controversies regarding its scientific validity, the name of Rankine to present day civil engineers is probably more closely associated with earth pressure theory than with any of his other activities.

Rankine the man

So much for Rankine and his scientific work. What do we know about Rankine himself from the writings of his contemporaries? His portrait and bust show him to have been of striking appearance. Professor Tait[24] says of him:

> 'Of the man himself it is not easy to speak in terms that, to the stranger, would not appear exaggerated. His appearance was striking and prepossessing in the extreme, and his courtesy resembled almost that of a gentleman of the old school. . . . His conversation was always interesting, and embraced, with equal ease, all topics however various. He had the still rarer qualification of being a good listener, also. The evident interest he took in all that was said to him had a most reassuring effect on the speaker, and he could turn without apparent mental effort from the prattle of young children to the most formidable statement of new results in physical science. . . . The questions which he asked on such occasions were almost startingly to the point, and he showed a rapidity of thought not often met with in minds of such calibre as his, where the mental inertia which enables them to overcome obstacles often prevents them being quickly set in motion'.

It might be thought that Rankine's activities as engineer, professor, and writer would have been sufficient to fill all his available time. However, in 1859 he accepted a commission as Captain in the Glasgow University Rifle Volunteers. He spent a month at the Hythe School of Musketry and on his return instructed the officers and sergeants of his Corps. In 1860 he was made Senior Major and commanded the second battalion of his regiment at the Volunteer Review held by Queen Victoria in Edinburgh. In 1864 he resigned his commission, finding, as he wrote, that it was 'impossible to attend at once to duties as field officer, and as professor, to engineering business, and to literary work'.

I have mentioned Rankine's interest at an early age in music. Apparently he had some skill at the piano, and he was in regular demand at University and other gatherings to accompany his singing of songs of his own composition. His fame in this connexion, as with his scientific work, had spread beyond Glasgow, and in 1871 when the British Association met in Edinburgh, he was hailed with universal acclaim as Lion-King, the Chief of the Red Lions, a group that Tait describes as the most unconventional of all clubs of scientific men. Their annual dinner appears to have been a hilarious occasion.

I have discussed his published texts. One further book I should mention is the volume on Rankine's *Songs and fables*[48] which was published after his death. As the preface to the book states, it is surprising to record that the qualities normally associated with a genius for philosophic research were not incompatible with the playful genial spirit which brightens the pages of the book. For many years I have never ceased to chuckle at Rankine's 'Mathematician in Love' (see Appendix). Another of his poems is the 'Three Foot Rule' which comments on what appears to have been a controversial issue in Rankine's time, dealing as it does with whether the metric system should be introduced in Britain. Apparently he wrote this for a Red Lion meeting in 1864.

Rankine was the most wonderful combination of the man of genius and of humour. How much more pleasant and effective is the contribution, scientific or otherwise, when you know behind it lies a man capable of having a twinkle in his eye.

You have done me a privilege in asking me to deliver this lecture, as through its preparation I have come to savour something of the aura and respect that this man generated among his contemporaries.

As Clerk Maxwell said, Rankine's death at the early age of 52 was as great a loss to the diffusion of science as to its advancement.

References

1. Barr A. W. J. Macquorn Rankine, a Centenary Address. *Proc. Roy. Phil. Soc. Glasgow*, 1920–22, **51**, p. 167.
2. Millar W. J. (Ed.) *Miscellaneous scientific papers by W. J. Macquorn Rankine.* Chas. Griffin and Co. London 1881.
3. Henderson J. B. *Oration on W. J. Macquorn Rankine. Jackson Wylie and Co.*, Glasgow, 1932.
4. Southwell R. V. W. J. M. Rankine: A commemorative lecture, *Proc. Instn Civ. Engrs*, 1956, **5**, Pt. I, 177–193.
5. Small J. The Institution's first President—W. J. M. Rankine. *Trans. Instn Engng Shipbldg, Scotland*, 1956–57, **100**, 687–697.
6. Cook G. Rankine and the theory of earth pressure. *Géotechnique*, 1950–51, **2**, 4, 271–279.
7. Todhunter I. and Pearson K. *A history of the theory of elasticity*, Vol. 2. Cambridge, University Press, 1893.
8. Timoshenko S. P. *History of strength of materials.* McGraw-Hill, New York, 1953.
9. Murray D. *Memories of the old College of Glasgow.* Jackson Wylie and Co., Glasgow, 1927.

10. RANKINE W. J. M. *Applied mechanics* C. Griffin and Co., London, 1858.
11. RANKINE W. J. M. *The steam engine and other prime movers* C. Griffin and Co., London, 1859.
12. RANKINE W. J. M. *Civil engineering.* C. Griffin and Co., London, 1862.
13. RANKINE W. J. M. *Machinery and millwork.* C. Griffin and Co., London, 1862.
14. RANKINE W. J. M. *Shipbuilding, theoretical and practical.* W. Mackenzie, London, 1866.
15. ROYAL SOCIETY. *Catalogue of scientific papers.* Vols 1–7 (1867–77) London, HMSO; Vol. 8 (1879) London, John Murray; Vols 9–19 (1891–1925), Cambridge, University Press.
16. RANKINE W. J. M. On the stability of loose earth. *Phil Trans. Roy. Soc.*, 1857, **147**, 9–27.
17. RANKINE W. J. M. Report on the design and construction of masonry dams. *The Engineer*, **33**, 1872, 5 Jan., 1–2.
18. PIPPARD A. J. S. Education and theory. *The Engineer*, 1956. Centenary number, 161–163.
19. PUGSLEY A. G. Scientific discovery. *The Engineer*, 1956, Centenary Number, 163–165.
20. THOMSON W., Lord KELVIN. Elements of a mathematical theory of elasticity. *Roy. Soc. Proc.* **8**, 1856–57, p. 85–87.
21. MAXWELL J. C. Tait's thermodynamics. *Nature*, **17**, 1878, 31 Jan., 257–259.
22. RANKINE W. J. M. On axes of elasticity and crystalline forms. *Phil. Trans. Roy. Soc.*, 1856, **146**, 261–286.
23. SMALL J. *Fortuna domus.* Glasgow, University Press, 1952.
24. TAIT P. G. Introduction to reference 2, Millar.
25. RANKINE W. J. M. On the resistance of ships. *Phil. Mag.*, 1858, **16**, 238–239.
26. RANKINE W. J. M. On the mechanical principles of the action of propellors. *Naval Arch. Trans.*, 1865, **6**, 13–39.
27. RANKINE W. J. M. On the causes of the unexpected breakage of the journals of railway axles; and on the means of preventing such accidents by observing the law of continuity in their construction. *Min. Proc. Instn Civ. Engrs*, 1843, **2**, 105–107.
28. RANKINE W. J. M. Description of a method of laying down railway curves on the ground. *Min. Proc. Instn Civ. Engrs*, 1843, **2**, 108–111.
29. RANKINE W. J. M. Description of Lieutenant D. Rankine's spring contractor. *Min. Proc. Instn Civ. Engrs*, 1843, **2**, 111–112.
30. WALKER J. Address of the President. *Min. Proc. Instn Civ. Engrs*, 1842–43, **2**, 21.
31. RANKINE W. J. M. *Introductory lecture on the science of the engineer.* London, C. Griffin and Co., 1857.
32. RANKINE W. J. M. Description of a safety drag, or apparatus for preventing accidents to trains ascending inclined planes, used on the Edinburgh and Dalkeith Railway since 1832. *Min. Proc. Instn Civ. Engrs*, 1844, **4**, 284–286.
33. RANKINE W. J. M. Account of the effect of the storm of the 6th of December 1847, on the four sea walls, or bulwarks, of different forms on the coast near Edinburgh: as illustrating the principles of the construction of sea defences. *Min. Proc. Instn Civ. Engrs*, 1848, **7**, 186–204.

34. Rankine W. J. M. and Thomson J. On the means of improving the Water Supply of Glasgow. G. Richardson, Glasgow, 1852.
35. Rankine W. J. M. and Thomson J. On the means of improving the Water Supply of Glasgow. Second letter, Glasgow, G. Richardson, 1852.
36. Weisbach J. *Principles of the mechanics of machinery and engineering.* Hippolyte Bailliere, London, 1848.
37. Gordon L. *Civil engineering and mechanics. Engineering aphorisms and memoranda.* A synopsis of lectures to be delivered in Session 1847–48. Schulze and Co., London, 1847.
38. Moseley. *The mechanical principles of engineering.* Longman, Brown, Green and Longmans, London, 1843.
39. Golder H. Q. Coulomb and earth pressure. *Géotechnique*, 1948–49, **1**, 1, 66–71.
40. Rankine W. J. M. On the mathematical theory of the stability of earthwork and masonry. *Proc. Roy. Soc.*, 1856–1857, **8**, 60–61.
41. Baker B. The actual lateral pressure of earthwork. *Min. Proc. Instn Civ. Engrs*, 1881, **65**, 140–186.
42. Wilson G. Some experiments on the conjugate pressures in fine sand, and their variation with the pressure of water. *Min. Proc. Instn Civ. Engrs*, 1902, **149**, 208–222.
43. Terzaghi K. A fundamental fallacy in earth pressure computations. *Jnl Boston Soc. Civ. Engrs*, 1936, **23**, 71–78.
44. Rankine W. J. M. *Useful rules and tables*, London, C. Griffin and Co., 1866.
45. Terzaghi K. and Peck R. B. *Soil mechanics in engineering practice.* John Wiley and Sons, New York, 2nd ed., 1967.
46. Pauker J. Dept. Ways of Communications, St Petersburg, 1889.
47. Kurdjumoff V. J. *Civiling*, 1892, **38**, 292–311, 1892.
48. Rankine W. J. M. *Songs and fables.* London, Macmillan and Co., 1874.

Appendix

The Mathematician in Love

I.

A Mathematician fell madly in love
With a lady, young, handsome, and charming:
By angles and ratios harmonic he strove
Her curves and proportions all faultless to prove,
As he scrawled hieroglyphics alarming.

II.

He measured with care, from the ends of a base,
The arcs which her features subtended:
Then he framed transcendental equations, to trace
The flowing outlines of her figure and face,
And thought the result very splendid.

III.

He studied (since music has charms for the fair)
The theory of fiddles and whistles,—
Then composed, by acoustic equations, an air,
Which, when 'twas performed, made the lady's long hair
Stand on end, like a porcupine's bristles.

IV.

The lady loved dancing:—he therefore applied,
To the polka and waltz, an equation;
But when to rotate on his axis he tried,
His centre of gravity swayed to one side,
And he fell, by the earth's gravitation.

V.

No doubts of the fate of his suit made him pause,
For he proved, to his own satisfaction,
That the fair one returned his affection;—"because,
"As every one knows, by mechanical laws,
"Re-action is equal to action."

VI.

"Let *x* denote beauty,—*y*, manners well-bred,—
"*z*, Fortune,—(this last is essential),—
"Let *L* stand for love"—our philosopher said,—
"Then *L* is a function of *x*, *y* and *z*,
"Of the kind which is known as potential."

VII.

"Now integrate *L* with respect to *d t*,
"(*t* Standing for time and persuasion);
"Then, between proper limits, 'tis easy to see,
"The definite integral *Marriage* must be:—
"(A very concise demonstration)."

VIII.

Said he—"If the wandering course of the moon
"By Algebra can be predicted,
"The female affections must yield to it soon"—
—But the lady ran off with a dashing dragoon,
And left him amazed and afflicted.

Equation referred to in stanza VI:

$$L=\phi(x, y, z)$$

$$=\iiint \frac{f(x, y, z)}{\sqrt{(\xi-x)^2+(\eta-y)^2+(\zeta-z)^2}}\, d\xi\, d\eta\, d\zeta$$

Equation referred to in stanza VII:

$$\int_{-\infty}^{+\infty} L\, dt = M$$